HEADStart

number

Shirley Clarke & Barry Silsby

Illustrated by Sascha Lipscomb

Headway • Hodder & Stoughton

NOTES FOR PARENTS

Research has shown that when children and parents work together at home, the child's work at school improves.

The purpose of the *Headstart* books is to provide activities which your child will enjoy doing and which will encourage learning to take place in the home.

Each of the activities is linked to one or more of the Attainment Targets of the Maths National Curriculum.

You can help your child get the most out of the book by

- *giving help* where necessary (for example, by reading instructions);
- *reading the advice below*. This gives further information and explains the purpose of each activity;
- *talking* to your child about an activity, to encourage him or her to put thoughts into words;
- showing enthusiasm and interest in your child's involvement in the book. (Confidence grows with adult approval.)

Pages 4–5 Ladybird, ladybird

There are two elements to this page: colouring in the ladybird border in alternate colours will help children look for repeating patterns (the basis of our number system); matching the ladybird twins will help focus on counting and same/different. If your child has difficulty, you could draw the missing ladybird for them and they could draw its spots.

Pages 8–9 Guess and find out

Talk to your child about the guess part of this, making sure they don't feel that the guess has to be right (children often worry about estimating).

If the real number is very different from your child's guess, praise them about how they have learnt something and how well they did it.

Pages 6–7 Number patterns

Once children can spot patterns like these they will remember their number bonds (for example, $5 + 5 = 10$, $3 + 2 = 5$) more easily. The focus of these pages is the patterns, rather than the individual sums.

Pages 10–11 Baby count

This activity can be used to talk about much more than counting. Talk about the different sizes of the animal families (we've tried to make them realistic) and compare them to human families. Talk about the differences between the creatures (fly/crawl, fur/feathers etc).

Pages 12–13 Calculator fun

Children should first feel confident with a calculator before they learn how to use it properly. Don't try to do 'sums' but let your child play with it, making up different numbers. After 'watch the calculator count' try pressing + 2 = = = or any other number and see what happens!

Pages 14–15 Money boxes

Use real money if your child needs help with this. Help them count the coins in the money boxes first. If your child chooses the wrong coin, say whether that would make the total too much or not enough (like a 'getting hotter' game) until they choose the correct coin by a process of elimination. You could then show them how this makes £1 using real coins.

Pages 16–17 Bubble game

Your child might find it easier to line up the dice e.g. 6 – 3, so that it is easier to subtract the smallest from the biggest.

If you cross off the bubbles with a coloured pencil, the game can be played again by using a different colour. Rolling dice many times in this way will help your child remember the number bonds.

Pages 18–19 Take half

'Half' is a difficult concept because we use it inaccurately. ('I'll be half a minute' usually means much more!) Talk about the different amounts on the stalls, saying whether the objects are more or less than half.

Pages 20–21 Happy birthday

This activity helps your child link the number symbol with counting (for example 5 is IIIII). See if they can space the candles round the cakes evenly by talking together about where they should go. The child's decorating could bring in repeating patterns again (red yellow, red yellow).

Pages 22–23 Calculator words

Children love finding 'words' on an upside down calculator. Although it is not the conventional use of a calculator, this kind of exploration will help your child become an inquisitive problem solver. See if they can make up some new words. (Remember the numbers have to be pressed in reverse order!)

Ladybird, ladybird

Carry on colouring in the ladybirds.

Make sure you keep the pattern going.

Join the ladybird twins.

Which ladybird is missing?

Draw it on the leaf.

Number patterns

Read the numbers down the columns.

Can you see the pattern?

Fill in the spaces.

1	+	9	=	10
2	+	8	=	10
3	+	7	=	10
	+	6	=	10
5	+		=	10
6	+		=	10
	+	3	=	10
	+	2	=	10
9	+		=	10
	+	0	=	10

Here is a different pattern.

Look down the columns again.

Can you fill in the spaces?

1	+	10	=	11
2	+	10	=	12
3	+	10	=	
	+	10	=	14
	+	10	=	15
6	+	10	=	
	+	10	=	17
8	+	10	=	
9	+	10	=	
	+	10	=	20

Guess and find out

How many toys can you hold without dropping any?

Guess [ ]

Now try it out!

I can hold [ ] toys.

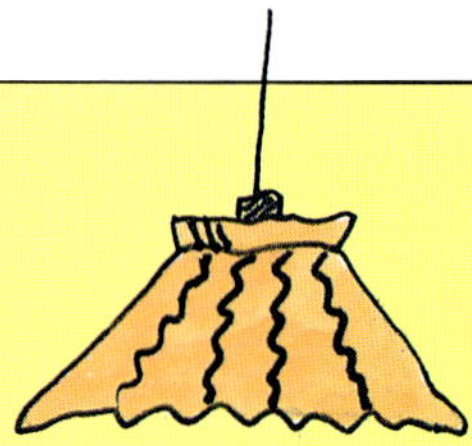

Are you taller than your mum if you stand on a chair?

Guess

Taller	
Shorter	
Same	

Now try it out!

I was [ ] than my mum.

How many times can you write your name in one minute?

Guess

Now try it out!

I wrote my name

times.

How long is your bedroom in pigeon steps? ( )

Guess

Now try it out!

My bedroom is

pigeon steps long.

Why not guess and find out some more things of your own?

Baby count

Can you help the mother count her babies?

Point to each one as you count or cross it off with a pencil.

Write how many babies there are in the circles.

Calculator fun

You need
A calculator

Can you make your calculator show
- your age?
- the age of someone in your family?
- the number 999999?

Think of other numbers to make on the calculator.

Watch the calculator count!

Press [+] [1] [=] [=]

Keep pressing [=]

Calculator numbers are made of lines.

Colour in these calculator numbers.

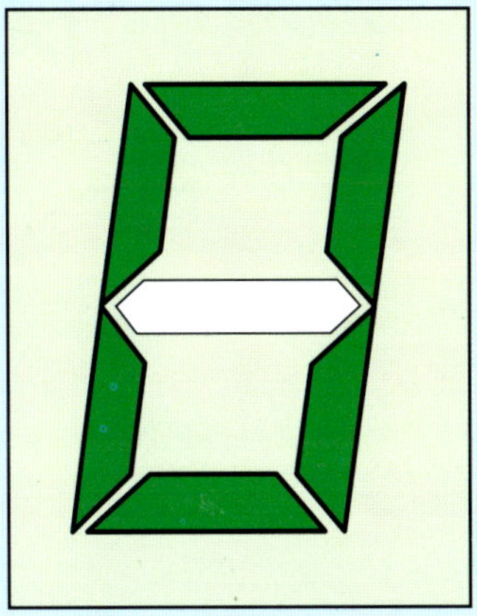

Money boxes

Ben needs £1 altogether in his moneybox.

What coin does he still need?

Circle the correct coin below.

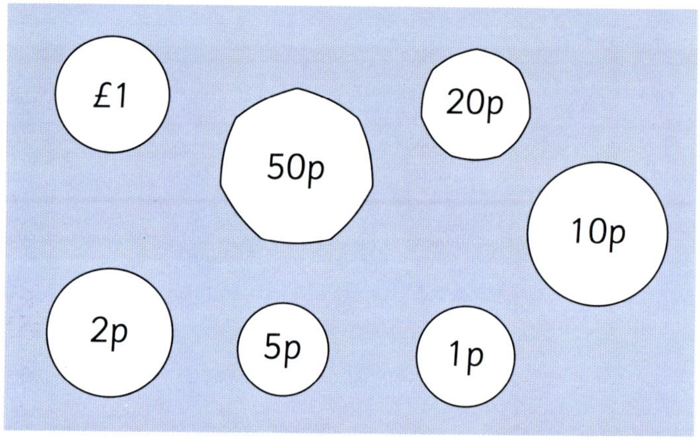

Tina needs £1 altogether in her moneybox.

What coin does she still need?

Circle the correct coin below.

Ali needs £1 altogether in his moneybox.

What coin does he still need?

Circle the correct coin in the box.

Anna needs £1 altogether in her moneybox.

What coin does she still need?

Circle the correct coin below.

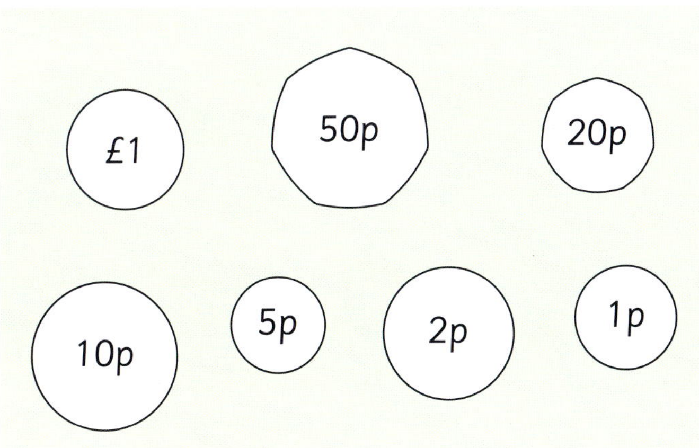

What do you think these children want to buy?
What would you buy with £1?

Bubble game

You need

2 dice

This is a game for two players.

1. Take it in turns to roll both dice.
2. Subtract the smaller number from the larger one.
3. Find the bubble with the answer on it and 'burst' it by crossing it through.
4. The winner is the first to burst all of his or her bubbles.

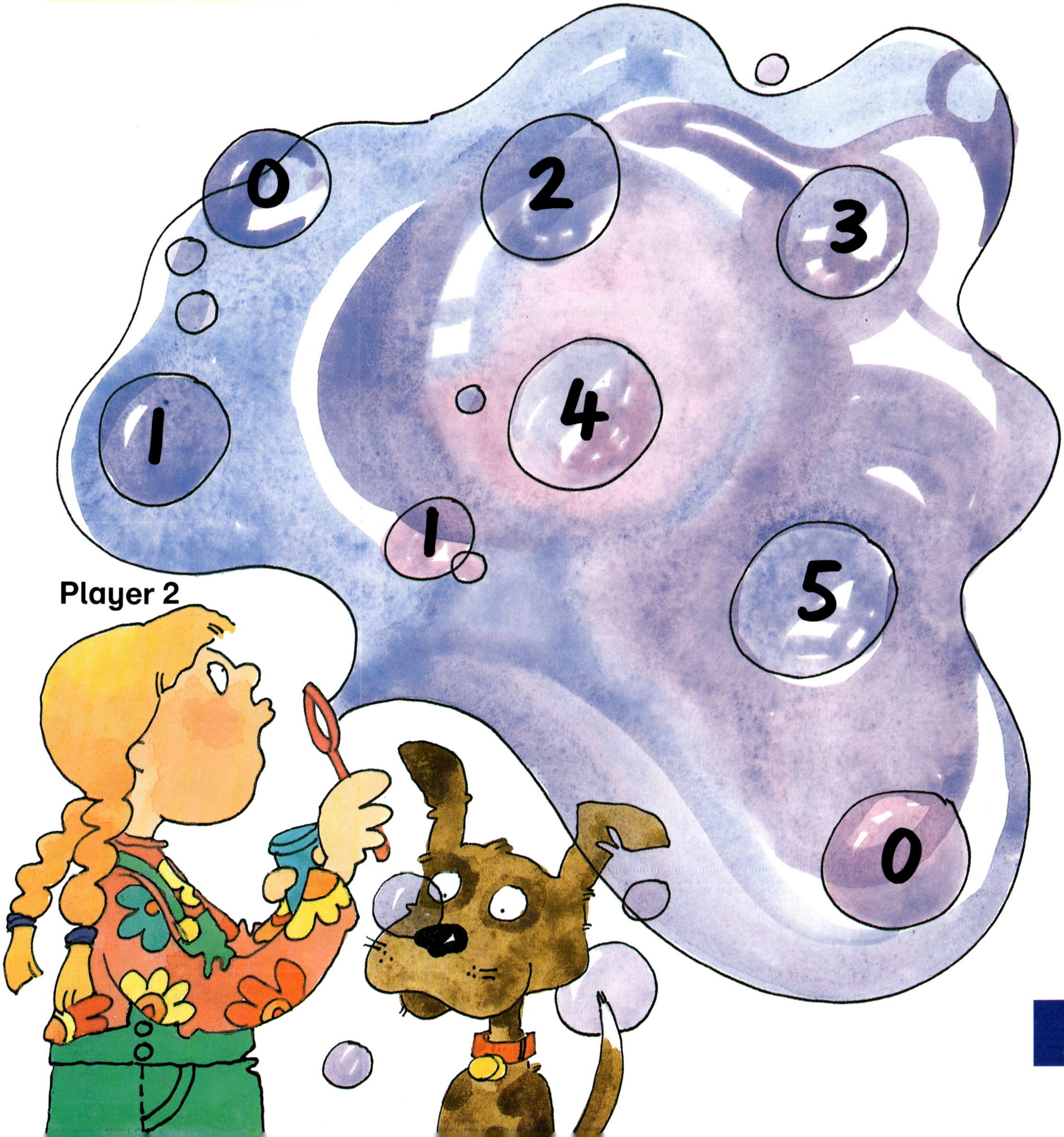

Take half

Put a ring round the right object on each stall.

MELON
FRESH FRUIT!
Half a melon please.
Sandwiches.
Just half a sandwich thank you.
Ham.
Cheese
Prawn.
Egg.

Happy birthday

Can you put the right number of candles on these children's birthday cakes?

Why don't you decorate the cakes as well?

8
6

Calculator words

Calculators can make words. Try this:

Press 3 0 4 5

Now turn the calculator upside down and read the display.

You need

A calculator

All of these numbers make words.

Each word is somewhere in the picture opposite.

Put a circle round each one.

The first has been done for you.

7 7 3 8

7 7 3 4 5

3 3 8

5 3 7 6 6 0 6

6 3 1

5 3 0 4 5

NUMBER 5–7 LINKS WITH THE NATIONAL CURRICULUM

ACTIVITY	USING AND APPLYING NUMBER	NUMBER AND NUMBER NOTATION	NUMBER OPERATIONS	ESTIMATING	PATTERNS IN NUMBER	MEASURES (LENGTH, WEIGHT ETC)
Ladybird, ladybird		✓			✓	
Number patterns	✓	✓	✓		✓	
Guess and find out	✓			✓		✓
Baby count		✓				
Calculator fun		✓				
Money boxes	✓		✓			✓
Bubble game	✓	✓	✓		✓	
Take half	✓		✓			
Happy birthday		✓				✓
Calculator words	✓	✓				

British Library Cataloguing in Publication Data
Clarke, Shirley
Headstart: number. – (Headstart)
I. Title II. Silsby, Barry III. Series
372.7

ISBN 0-340-54455-4

First published 1991
Second impression 1992

Typeset by Oxprint Limited, Oxford OX2 6TR

Printed for the educational publishing division of Hodder & Stoughton Ltd, Mill Road, Dunton Green, Sevenoaks, Kent by Colorcraft, Hong Kong.